Ludvigsen Library Series

CUNNINGHAM SPORTS CARS

American Racing Legends 1951-1955

Introduction by
Karl Ludvigsen

Iconografix

Iconografix
PO Box 446
Hudson, Wisconsin 54016 USA

The information in this book is true and complete to the best of our knowledge. All recommendations are made without any guarantee on the part of the author or Publisher, who also disclaim any liability incurred in connection with the use of this data or specific details.

We acknowledge that certain words, such as model names and designations, mentioned herein are the property of the trademark holder. We use them for purposes of identification only. This is not an official publication.

Iconografix books are offered at a discount when sold in quantity for promotional use. Businesses or organizations seeking details should write to the Marketing Department, Iconografix, at the above address.

Library of Congress Control Number: 2003109097

ISBN 1-58388-109-3

03 04 05 06 07 08 09 5 4 3 2 1

Printed in China

Cover and book design by Dan Perry

Copyediting by Suzie Helberg

COVER PHOTO: Crouching as menacingly today as it ever did in its prime, this Cunningham C-4R was photographed for the cover by Art Eastman. It represents the quintessence of the success of the short-lived Cunningham marque.

BOOK PROPOSALS

Iconografix is a publishing company specializing in books for transportation enthusiasts. We publish in a number of different areas, including Automobiles, Auto Racing, Buses, Construction Equipment, Emergency Equipment, Farming Equipment, Railroads & Trucks. The Iconografix imprint is constantly growing and expanding into new subject areas.

Authors, editors, and knowledgeable enthusiasts in the field of transportation history are invited to contact the Editorial Department at Iconografix, Inc., PO Box 446, Hudson, WI 54016.

THE MAGNIFICENT CUNNINGHAMS

by Karl Ludvigsen

In retrospect it's no easy task to communicate what the Cunningham sports cars meant to car enthusiasts in America in the mid-20th Century. America's post-war autos were worthy enough, but for sporting—not at all. Not since the Auburn Speedsters and Cord 810s of the mid-1930s had American automakers produced cars that appealed to the person for whom driving was more than a way of trundling from point A to point B. Thus in 1951 the car-enthusiast world was galvanized by the news that the B. S. Cunningham Company of West Palm Beach, Florida would not only produce a new American sports car but would also race a team of three cars at Le Mans!

This was big news not only in America but also abroad. Briggs Cunningham made many friends at Le Mans in 1950 with a sortie by two Cadillacs that placed 10th and 11th overall. That he would now build his own cars and race them in the world's best-known sports-car contest was a sensation of the first magnitude. The white and blue Cunningham cars were destined to compete officially from 1951 through 1954. They scored 14 race victories, all of them in America. To qualify to race at Le Mans, Cunningham also produced a handful of road cars that now rank as choice collectors' items.

Who was this fellow Cunningham? Above all, he was fortunate in having the means to pay his own way to Le Mans, for money problems rarely troubled him. Briggs was born in Cincinnati in 1907 of parents who were both seriously wealthy. Cars always appealed to the young Cunningham. In the early 1930s he competed in some of the events organized by the Automobile Racing Club of America. A friend remembered, "Briggs and I had many discussions about racing together in Europe or importing the European style of road racing to this country in the 1932-1933 period." Also promoting this idea were the Collier brothers, Miles and Sam, both of whom raced in Europe during the 1930s.

His sports-car enthusiasms re-ignited after the war; in the autumn of 1949 Cunningham began to talk about taking a team of American cars to Le Mans. "I knew the Collier boys," he recalled, "and Miles had been over there in an MG before the war. We got talking, like you do, and someone said, 'Gee, why don't we put together a car?'" After the exploratory Cadillac campaign in 1950, work began in earnest on the first Cunningham. Its prototype was a Cadillac-powered Healey Silverstone built by Long Island-based Frick-Tappett Motors. At this racing and specialty garage Bill Frick exploited his talent for hot-rodding and preparing midgets raced successfully by "Ted Tappet," the nom de course of Phil Walters.

Cunningham merged his interests with Frick-Tappett Motors and moved it to West Palm Beach, Florida, where the B. S. Cunningham Company was established in September 1950. The firm's first product was a Cadillac-powered prototype called the C-1. From this evolved the C-2. The latter's heart was the hemi-head Chrysler V-8 that had just been announced for that company's 1951 models. Then the unchallenged leader in the budding horse-power race, its standard output was 180 bhp at 4,000 rpm. With an increase in compression ratio from 7.5 to 8.6 to one and a log manifold that carried four downdraft Zenith carburetors, power was raised to 220 bhp for Le Mans.

The C-2's robust foundation was a chrome-molybdenum-steel frame of 3-inch tubing for its side rails, cruciform center bracing and rear crossmember. A Ford crossmember for its Ford parallel-wishbone front suspension was welded into the front end of the frame. Additional stiffening and body support were provided by a superstructure of 1¼-inch tubing along the cowl, sills and fender wells. Rear suspension was de Dion with a tubular axle passing behind the differential, sprung by coils. Eleven-inch Cadillac brakes were mounted inboard at the rear.

3

The team of three racing C-2s was assembled from March through May 1951 by a 40-man work force putting in some 80-hour weeks. In the Le Mans race two were eliminated by malfunctions aggravated by rainstorms. The remaining Cunningham driven by John Fitch and Phil Walters worked its way up to second place at 5:00 A.M. and held that position until 10:00 A.M., though some eight laps behind the leader. "People were very impressed," said Fitch, "when this big white two-ton sports car roared past the pits and under the Dunlop Bridge at very high speed!" His C-2 was timed at 152 mph on the Mulsanne Straight.

Slowed in the last five hours by burned valves and weary connecting-rod bearings, the Cunningham C-2 placed 18th overall and first in its class, ahead of an equally fatigued Cadillac-Allard. Nothing to wave the Stars and Stripes about, but those six hours in second place, by an untried design, made a major impression at Le Mans. The three team Cunninghams redeemed themselves in their native country in the two main sports-car races of 1951, Elkhart Lake and Watkins Glen. They placed first and sixth at Elkhart and first, second and fourth at the Glen.

On a platform much like that of the C-2, but with a live rear axle instead of the de Dion, Briggs and his Florida crew planned the C-3 production model that would qualify them for future Le Mans entries as a bona fide manufacturer. Only one C-3, a coupe, was built entirely at the Cunningham factory. Subsequent production cars were bodied in Turin by Alfredo Vignale. Their subtle fastback coupe design was by Giovanni Michelotti.

Though the first Vignale C-3 had a 105-inch wheelbase, the same as the C-2, later cars had a 107-inch wheelbase to provide a passable 2+2 seating condition. The only Cunningham modifications to the massive V-8s supplied by the Chrysler Industrial Division were external, in the form of two-inch dual exhausts with Porter mufflers and the Cunningham log manifold fitted with four Zenith downdrafts, topped by Hellings wire-mesh air cleaners. Of the total output of 23 Vignale-bodied Cunninghams, five were to be handsome convertibles.

In parallel with its creation of the C-3 the little Cunningham outfit created an all-new car for racing in 1952. Designated C-4R, this became the most durable and successful competition Cunningham and the model most widely known around the world. Slimmer by six inches, shorter by 16 inches and lighter by 990 pounds than the C-2, the C-4R was intended to be faster and better. In all respects it succeeded. It benefited from the design talents of veteran engineer G. Briggs Weaver, who joined Cunningham in October 1951.

Carried over to the C-4R was the C-2's Chrysler engine, which with a roller-tappet camshaft now developed 325 bhp at 5,200 rpm. Also retained was the Ford front suspension, integrated with an entirely new frame whose side members were fabricated from two superimposed steel tubes, joined by sheet-steel side gussets and severely kicked up at the rear. Instead of a de Dion rear suspension a live Chrysler rear axle was guided by parallel trailing arms and a lateral Panhard rod, using geometry worked out by Chrysler engineers.

The prototype C-4R roadster was first shown on March 12, 1952. It was joined by a sister roadster and a coupe version, the C-4RK. The "K" stood for "Kamm" in recognition of Dr. Wunibald I. E. Kamm, German aerodynamicist who had come to West Palm Beach to suggest altering the coupe's clay model to the truncated tail shape with which he was associated. As executed by the Cunningham bodymakers—who gave it a flat, slit-like windshield, huge fuel filler in the low roof, twin scoops on the cowl and numerous louvers—this was one of the meanest-looking autos ever to take the road.

At Le Mans, Phil Walters stormed the C-4RK into a first-lap lead, then turned the fastest 1952 lap for the Cunningham team on his second round at 105.6 mph. Walters held third until the first change of drivers, after which his co-driver put the coupe into a sand bank for a stay of one hour and forty minutes. It retired on lap 66 with the same valve-gear trouble that stopped one of the roadsters, provoked by over-revving on downshifts. This was caused by both brake problems and the continued use of three-speed gearboxes when a new five-speed design failed in practice. Briggs placed fourth overall in the remaining roadster after an iron-man driving stint of 19½ hours.

The C-4Rs returned to Le Mans in 1953 and 1954. They finished seventh and tenth in 1953 and third and fifth in 1954. In domestic competition they won in 1952 at Allentown, Thompson, Elkhart Lake (in the last race on the original circuit) and at Albany, Georgia. Walters and Fitch co-drove to victory at Sebring in 1953 in the first-ever race for the sports-car manufacturers' championship. While in Europe that year, Briggs Cunningham and Sherwood Johnston placed third in the Reims 12-hour race. Victories at Thompson, Floyd Bennett Field, Albany and the big "East versus West" race at March Field near Riverside, California wound up 1953.

According to Phil Walters, 1953 was the year Cunningham should have won Le Mans, if a Cunningham car were ever to win it. The "if" car was the new C-5R, which had a straight tubular front axle guided by very long parallel leading arms on both sides and a vertical slide at the center. Long parallel trailing arms located the live rear axle, and all four corners were sprung by longitudinal torsion bars. Slung under the axles was a narrow, straight frame of twin oval steel tubes. Walters and Briggs Weaver gave the C-5R huge 17-inch drum brakes offset inboard from its wheels. Powered by the big Chrysler engine, rated at 310 bhp in 1953, at Le Mans it registered the fastest timed speed of all the entries: 154.81 mph compared to 146.21 mph for the C-4R roadster and 150.24 for the C-4RK coupe.

The prognosis was promising, but the C-5R couldn't hold the new Weber-carbureted, disc-braked C-Type Jaguars. As Phil Walters said, "We could out-speed and out-drive the Jags, but on their brakes alone they easily cruised faster than we could. We ran the first two hours at our target speed, which was 104, and then raised our sights to catch the Jaguars. We couldn't do it." Walters and John Fitch finished right on their target speed (plus 0.026 mph), but they were seven laps behind the winning Jaguar. The year 1953 marked the first time that every Cunningham finished the race with all three placed among the first 10.

No all-new Cunninghams were built for Le Mans in 1954. Talks with Mercury Marine's Carl Kiekhaefer about a new two-stroke V-12 engine were in train, so until that was ready it was decided to use the 4.5-liter Ferrari V-12 in the next car. Briggs Weaver designed the C-6R around this engine. Time ran out before the car could be completed, so a Ferrari 4.5 chassis with live rear axle was rebodied by the West Palm Beach crew with an oval grille and equipped with liquid-cooled drum brakes. At Le Mans this Cunningham-Ferrari retired halfway through the race with rear-axle failure after being slowed by a broken rocker arm.

Meanwhile Briggs Weaver was making progress with the C-6R chassis. Its frame had large twin tubes on each side forming a pyramidal truss structure that started from a big tubular crossmember at the front. The structure peaked at a cowl cross-member and then tapered away to the intricate rear end. Front suspension on the C-6R was by short and long wishbones, hollow fabricated parts of ineffable artistry. The rear suspension was by coil-sprung de Dion axle, guided by parallel trailing arms and a vertical channel. "By the time we built the C-6R," Briggs Weaver recalled, "we had the best crew I ever worked with. I could draw up anything, take it out to the shop, and they could make it. They did beautiful work."

To power his C-6R, Cunningham bought an "Offy" engine from Meyer & Drake, a twin-cam, 16-valve destroked 220-cubic-inch four that measured 100 x 92 mm for a displacement of 2,942 cc. Put on the dynamometer in West Palm Beach in late October 1954, this was developed with the frequent telephone cooperation of Leo Goossen of Meyer & Drake. Bob Blake and Herbert "Bud" Unger built a curvaceous body on the 650-pound chassis, bringing the total dry weight of the C-6R to a commendably light 1,904 pounds. After an inconclusive entry at Sebring, Briggs Cunningham and Sherwood Johnston shared the three-spoke wheel of the C-6R in 1955's Le Mans race. An early and unexpected setback was the loss of two of the indirect ratios in its four-speed ZF box. Finally they had only top gear left. Running under these conditions provoked a burned piston, which retired the car at 10:20 A.M. after 202 laps, while it was running 13th.

After Le Mans Briggs drove the C-6R in the inaugural race on September 11 at Road America (Elkhart Lake), where the Meyer-Drake retired itself from sports-car racing with explosive finality. By then Cunningham was committed to a new racing program with D-Type Jaguars so his C-6R was set aside until the approach of Sebring in 1957. For that race a D-Type engine, radiator and gearbox were installed, but pre-race problems caused it to be scratched. It was later entered in SCCA events in which it performed well because it handled superbly and was lighter than the D-Types, but the team's emphasis was necessarily on the all-Jaguar machines.

Thus did the last of the Cunninghams retire, reluctantly and with a grace worthy of the name, from active competition. It represented the ultimate extrapolation of the all-American Cunningham design concept as well as a full stop to the career of Briggs Cunningham as a builder of sports-racing cars. As he explained to a reporter, "I tried to win the Le Mans 24-hour race for Americans with an American car. I have stopped trying because I, personally, can't afford to compete against the biggest manufacturers in Europe, especially when I have to build my cars from the smallest fitting on up." Surprisingly little has changed in the subsequent half-century.

During Le Mans preparations, Derek Waller pictured a typical Briggs Cunningham gesture. Waller belonged to a British team assisting with timing and scoring, another member of which was Geoffrey Kramer, here checking the Le Mans regulations. Ace mechanic and car preparer Alfred Momo is just visible behind Kramer.

As was his right and privilege, Briggs Cunningham cornered the market in top-ranked American racing drivers for his campaigns at home and at Le Mans. His two aces were Phil Walters, left, and John Fitch. Walters also served as general manager of Cunningham's car-building company in West Palm Beach, Florida.

Rumors in American racing circles about Cunningham's car-building plans were confirmed on April 20, 1951, when his company released the first photos and description of the car it was planning to race at Le Mans that year and later to manufacture for sale. Although otherwise boldly all-American, it wore Italian Borrani wire wheels.

In its frontal aspect the first Cunningham—largely engineered by Bill Frick—showed a certain debt to the designs of Italy's Carrozzeria Touring for Ferrari. Grilled apertures were provided for air to cool its front brakes.

The first Cunningham was not a small automobile. Nevertheless its designers, working from a small clay model, succeeded in giving its rear view a certain elegance. Twin tail pipes served the first car's Cadillac engine.

Photographed among Florida's palms, the first Cunningham had a small hood bulge and a curved windshield. Commented Britain's *The Motor*, "Its handsome and unadorned lines are particularly noteworthy as coming from a country where a typical automotive product is heavily overlaid with painted or plated ornamental hardware."

Deep-sided and adjustable, the individual bucket seats met the need for excellent lateral support for driver and passenger. These, commented Tom McCahill in *Mechanix Illustrated*, "Custom-built by a master upholsterer with the finest real leather you can buy, would set you back about $600 a pair if you wanted a couple for your living room."

This first Cunningham prototype, retrospectively designated the C-1, featured instrumentation that included fuel pressure and oil temperature. This car was used in competition only once, in the Mount Equinox hillclimb in October 1951. Only George Weaver's Grand Prix Maserati bettered John Fitch's time over the full course.

Rear suspension of the first Cunninghams was de Dion type, with the axle tube connecting its two rear hubs visible at the left. A trailing radius arm, extending to the right, located each hub. A flexible joint in the center of the de Dion tube allowed the hubs to rise and fall individually without being constrained by the axle's torsion.

Seen under construction at Cunningham's factory, his car's frame relied on a massive cross-braced tubular platform on which a superstructure braced the cowl and supported the body. A de Dion axle rested on the table behind it.

The first Cunningham's 11-inch rear drum brakes were mounted inboard adjacent to its differential and final drive. Made by Pat Warren, the latter incorporated an extension containing quick-change gearing that was modified, on some of the Le Mans cars, to provide a two-speed rear axle to supplement the three-speed manual transmission.

Brazing copper strips around the drum's periphery enhanced cooling of the 1951 Cunningham's 12-inch front drum brakes. Helpful though this was, braking would not be a strong point of the first Cunninghams.

Displacing 331 cubic inches (5,424 cc), the Chrysler V-8 engine used by Cunningham was fed by log manifolds carrying four Zenith carburetors. In Le Mans tune it was credited with 220 bhp and capable of 5,500 rpm.

The three Cunninghams built to compete at Le Mans in 1951 were designated the C-2 model. Painted white with blue striping, they proudly carried the Cunningham checkered-flag badge showing their country of origin.

Before they were shipped to France on the *Mauretania*, testing of the three team cars had consisted of only four, five and eleven miles respectively on Florida's highways. The weather in France came as a sharp contrast, the squally rain during June 23rd and 24th being considered the worst at Le Mans since the 24-hour race was first run in 1923.

Briggs Cunningham's expression reflected the fact that he and his fellow drivers enjoyed only limited practice time in 1951, frequent interruptions being needed to attend to the still-new C-2s. The hood, with its cooling vent, was leather-strapped as required by the Le Mans rules. A low speedboat-style windshield was adopted for the race.

At the start in 1951 Phil Walters was already away in Cunningham number 4. Briggs was next in number 3, followed by the number 5 C-2 of George Rand, driving, and Fred Wacker. Number 14 was an aerodynamic Bentley.

Alongside two of the Cunninghams at the start was one of the quick new C-Type Jaguars that came as a distinct shock, both to the Cunningham team and to other competitors. This one went on to win the 1951 24-hour race.

The C-2 Cunningham was ill suited to the wet conditions of the 1951 race. Experienced racer Fred Wacker called it "a large, heavy, powerful car that had certain shortcomings in that the steering was hard and the general character of the car was somewhat truck-like. In short, it was a bit of a beast although it was, certainly, a noble effort."

In 1951 Cunninghams number 3 and 5 were both eliminated by crashes in Le Mans's difficult conditions. "It would not, I think, be an easy car to extract from an emergency," said Britain's Laurence Pomeroy, Jr., after trying the Cadillac-powered prototype, and so it proved for both George Rand and Briggs's co-driver George Huntoon.

Driving number 4, Phil Walters and John Fitch moved up to second place just past the halfway mark and held second through the 18th hour. Through 20 hours it was still lapping as fast as had the previous year's winner.

Engine problems hit the Fitch/Walters C-2 just before noon on its final day. It struggled to the four o'clock finish, placing 18th and winning its class. Though blunted, Cunningham's Le Mans campaign had shown its potential.

Getting in some unofficial practice for the Watkins Glen, New York Grand Prix on September 15, 1951, Briggs Cunningham prepared to set out on the New York resort town's main street with a passenger. His team car continued to use Borrani wire wheels while the other Cunninghams were converted to Halibrand magnesium wheels.

Here chased by a Cadillac-Allard, Briggs Cunningham started out in seventh place in the 99-mile Grand Prix at Watkins Glen and moved up to his fourth-place finishing position by the tenth of the 15 laps. His C-2s raced in America with the full two-piece windshields with which they would later be sold to the public.

Karl Ludvigsen caught Phil Walters giving a lady passenger an exciting ride around the challenging 6.6-mile Watkins Glen road course during practice for the 1951 Grand Prix. Compared to the C-1 prototype, the C-2 model had sprouted air inlets to cool its rear brakes and a substantial hood bulge with an air entry to feed its carburetors.

For the American races the Chrysler engine's power was increased to 270 bhp at 5,500 rpm with changes to both cam timing and porting. At Watkins Glen John Fitch finished second while Phil Walters, shown, won with a margin of one and a half minutes at a speed of 77.65 mph. The first four finishers bettered 1950's winning average.

The team's three C-2 Cunninghams were refitted as road cars and sold through International Motors of Los Angeles. All were fitted with Borrani wire wheels. This car later received a rollover bar and external exhausts.

As converted for public sale the three C-2s had new egg-crate grilles with a "C" emblem in the center. They were also fitted with robust bumpers at both front and rear, an amenity that this C-2 has later omitted.

The owner of this Cunningham C-2 took no chances on starving its Chrysler engine of fuel. He fed it through six twin-throat downdraft racing carburetors on a special manifold.

Irving W. Robbins, Jr., of Palo Alto, California raced his black C-2 Cunningham on several occasions. He drove it at Palm Springs on March 23, 1952, placing poorly. On September 7, 1952, shown, Robbins was 10th overall in the main race at Elkhart Lake, Wisconsin. Phil Hill also briefly raced a C-2 on the West Coast.

Bodied in aluminum like the C-2 models, a coupe version was produced by Cunningham's craftsmen in West Palm Beach and dubbed the C-3. This was intended as the basis of the road-car production that would qualify Cunningham as a car manufacturer and thus enable him to continue entering prototypes at Le Mans.

Seen under construction in Florida, the C-3 coupe proved far too costly to manufacture under American conditions. The single coupe produced in this manner was bought by Carl Kiekhaefer, owner of Mercury Marine.

For its production models the B. S. Cunningham Company contracted with Turin's Alfredo Vignale to manufacture bodies on chassis shipped from Florida. Vignale created a handsome interior with the four subsidiary instruments on the left sharing space with a clock. Center armrests folded down to give better lateral bracing.

Styling of the C-3 Cunningham Continental coupe was based on a sketch by Giovanni Michelotti, whose concept lent itself to three-tone paint schemes, as on this car shown at the Paris Salon in October 1952.

Carburetion akin to that of the 1951 Le Mans cars fed fuel to the C-3, with four Zeniths topped by Hellings wire-mesh air cleaners. C-3 power was quoted at 220 bhp at 4,000 rpm in 1952 and 235 bhp at 4,400 rpm in 1953.

Seen being serviced in the Queens, New York workshops of Alfred Momo, a C-3 Cunningham revealed its Ford-based front suspension with 11-inch Mercury brakes. A coil-sprung Chrysler rear axle was used.

During its lifetime this C-3 Cunningham acquired non-original bumpers. While the first Vignale-built car had a 105-inch wheelbase, the final dimension was 107 inches to provide an approximation of occasional rear seating.

For the early 1950s the Cunningham coupe was a handsome machine. In the style of the day, of course, it wouldn't do to be seen without whitewall tires. This car, photographed in 1960, was wearing slimmer whitewalls.

At the Geneva Salon in March 1953 the first convertible Cunningham C-3 was unveiled. After completion in Turin the Vignale-bodied cars were shipped from Genoa for final completion at West Palm Beach.

With its top up the Cunningham C-3 Continental convertible was as handsome, in its own way, as the coupe. Vignale required two months to fit its body to each of the chassis it was supplied by West Palm Beach.

Most C-3s—like this convertible—were delivered with Chrysler's Fluid-Torque semi-automatic transmission. Supplementary engine braking could be engaged by a pushbutton on its column-mounted gear selector.

On the C-3 Cunninghams, 16-inch wheels were standard although cars could also be delivered with larger-section tires on 15-inch rims. Michelotti's design provided these sports cars with a distinctive and attractive identity.

Built as it was around the big 750-pound hemispherical-head Chrysler V-8, no Cunningham was going to be small. However, for 1952 every key dimension of the new C-4R racing model was significantly smaller. Astonishingly, at 2,410 pounds dry it weighed almost half a ton less than its C-2 predecessor.

Carrying the "R" suffix to show its dedication to racing, the C-4R was the first Cunningham to be professionally engineered with the help of veteran (du Pont cars and Indian motorcycles) G. Briggs Weaver, who had joined the Florida-based team. Lower corners of the new car's nose were cut away to deliver more cooling air to its brakes.

Ford suspension parts continued to be used in the C-4R's front end. With fine lateral finning, as recommended by Chrysler, its brake drums proved to be structurally weak at Le Mans and suffered from cracking.

Chrysler's brake-cooling concept required special magnesium wheels with an aluminum cover that made the wheel function like a centrifugal blower to draw air over the finned brake drums. Drum diameter was 13 inches.

At Le Mans in 1952 the three Cunninghams lined up for the start at the head of the field by virtue of their cyl-inder capacity. Only a supercharged French Talbot outranked them. Drivers checked their cars in preparation for the start.

John Fitch and George Rice handled C-4R number 3 at Le Mans in 1952. They held an excellent third place in the fourth hour but by the sixth had retired with valve-train problems. These were provoked by having to rely on Cadillac transmissions with only three forward ratios, after their five-speed gearboxes proved faulty in practice.

A new feature of the 1952 racing Cunninghams was a steering wheel sloped at approximately 45 degrees. Akin to American Indy-car practice, this was felt superior for long-distance racing because it allowed drivers to use their shoulders as well as their arms to steer. All the team's drivers adapted well to it without demur.

In 1952 Briggs Cunningham, concerned that co-driver Bill Spear might overtax a slipping clutch and aware as well that Spear's vision left something to be desired, stayed at the wheel of his C-4R for almost 20 of the 24 hours. Briggs was rewarded with fourth place in the only one of his Cunninghams that finished the race.

In 1952 Briggs roared past the pits in his C-4R, which developed 325 bhp at 5,200 rpm with the help of a new roller-tappet camshaft. Chrysler Engineering took a close interest in Briggs's program and was supportive of its efforts. The Cunninghams were the inspiration for Chrysler's 300-series high-performance production cars.

In front of the main stands at Le Mans in 1952 the Cunningham C-4R showed its slimmer lines. On a wheelbase of 100 instead of 105 inches, it was 16 inches shorter and 6 inches narrower than its predecessor. Track width was reduced from 58 inches to 54. Nevertheless room was still found for 50 gallons of fuel on board.

Though no longer the Cunningham team's front-line effort, the C-4R returned to Le Mans in 1953. The lone road-ster entered was driven by Briggs himself along with Bill Spear. They were timed at 146.21 mph on the Straight.

On June 14, 1953, Briggs acknowledged the flag at the end of the Le Mans race. He and Bill Spear placed seventh overall at an average of 101.143 mph, covering 305 miles more than they had the previous year.

In 1953 the C-4R Cunninghams used conventional finning for the iron braking surfaces of their drums. Visible were the worm-and-roller steering gear, adjustable Houdaille shock absorbers and front anti-roll bar.

A new amenity on the 1954 C-4Rs at Le Mans was a vertical-column decelerometer. This helped the drivers pace their braking more precisely. Cylindrical oil coolers were mounted on the cowls of the C-4R roadsters.

For 1954 the C-4R engines were fed by four twin-throat Solex carburetors. This gave each cylinder of the Chrysler V-8 its own venturi and ram tuning to improve mid-range torque. Although four-speed Siata gearboxes had been used successfully in sprint races, for Le Mans the team continued to rely on three-speed Cadillac transmissions.

Modified Chrysler engines for the Cunninghams were built by the experienced Eddie Bourguignon in Florida. Speeds over the timed kilometer at Le Mans in 1954 for the two C-4R roadsters were 148.05 and 147.56 mph.

Two tall drivers, Bill Spear and Sherwood Johnston, drove C-4R number 2 at Le Mans in 1954. Captured by Rudy Mailander taking the fast bends at White House, they seized third at the 15th hour and held it to the finish.

The Spear/Johnston Cunningham passed the wreckage of an Aston Martin DB3S coupe. Driving the other C-4R in 1954, John Gordon Benett remarked that he was "so glad I was driving a Cunningham" when he encountered the Aston Martin's fresh debris on the track. As well as being fast, the C-4Rs were gratifyingly rugged.

Difficult conditions during the rain-swept 1954 Le Mans race kept the Cunninghams from equaling their best distance achievement of 1953. The number 1 car, driven by Briggs Cunningham and John Gordon Benett, placed fifth overall at an average speed of 95.374 mph. It had held that position for the final 10 hours of the race.

The Spear/Johnston Cunningham was given the checkered flag at the end of its run to third place at Le Mans in 1954. Its average for the race was 98.640 mph. With the advancing pace of their rivals the C-4Rs were no longer the threat for overall victory that they'd been in 1952, but their 24-hour reliability was now impeccable.

Sportingly, Briggs Cunningham also entered his cars in races throughout the United States. In the main event at Elkhart Lake, Wisconsin on September 7, 1952, Briggs placed third in a one-two-three Cunningham sweep.

John Fitch piloted the winning Cunningham C-4R in the Elkhart Lake Cup race of 1952, averaging 87.5 mph over 201.5 miles on roads near this Wisconsin resort town. His only rival was another Cunningham driven by Phil Walters.

As entered in American races the cockpit of the C-4R was workmanlike to a fault. For U.S. racing the Siata four-speed transmission was used and braking was aided by a Bendix vacuum booster, first raced at Le Mans in 1954.

John Fitch was relaxed at the wheel of his winning C-4R at Elkhart Lake in 1952. The three Cunninghams defeated two C-Type Jaguars making their American racing debuts. At Le Mans, however, the shoe was on the other foot.

The most awesome Cunningham was the C-4RK coupe first raced at Le Mans in 1952. It was exhibited at Bridge-hampton on May 24, 1952, when the C-4R first raced and led until it was black-flagged for a loose exhaust pipe.

Expected to benefit from a lower drag coefficient, the coupe was expressly designed with Le Mans in mind. Relevant as well was that the drivers of the open C-2s in 1951 had complained of being badly wind-whipped.

Unlike its roadster sisters, the C-4RK had a one-piece aluminum hood that hinged up and forward to give complete access to its engine and front suspension. With its four downdraft Zenith carburetors, the hemi-head Chrysler V-8 engine was similar to those powering the C-4R roadsters, with smoothly blended exhaust manifolding.

The coupe's "K" designation was a bow to German aerodynamicist Wunibald Kamm, who had consulted on its design. In accord with drag-reduction principles that he had developed before the war, Kamm recommended that the coupe's tail be sharply truncated instead of extending to a point as its designers had originally proposed.

Seen just before the end of the first hour at Le Mans in 1952, the C-4RK in the hands of Phil Walters set its fastest lap—and the fastest of the Cunninghams that year—on its second lap of the circuit at 105.6 mph. On the race's first lap the big coupe roared past the pits in the lead, always a memorable achievement at Le Mans.

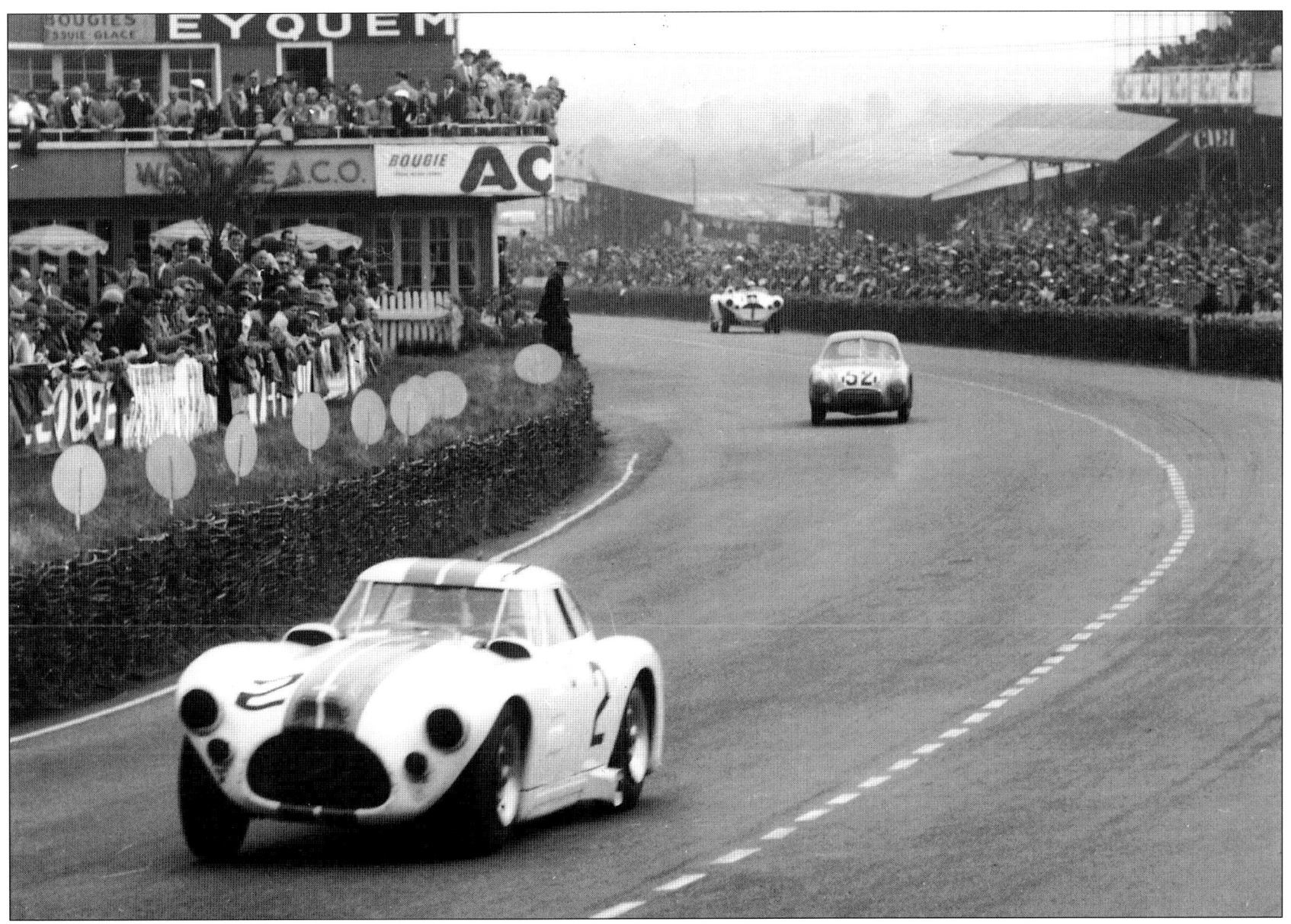

At the end of the third hour in 1952 the coupe was running second to a fragile French Gordini and thus looked poised for victory. After taking over from Phil Walters, however, Indy racer Duane Carter put the C-4RK into a sandbank and needed an hour and a half to extricate it. During the sixth hour its valve gear cried enough.

From 1953 onward the C-4RK played a back-up role at Le Mans. Driven to finish by John Gordon Benett and Charles Moran in 1953, it was here heading one of the fast Alfa 3½-liter Romeo coupes that failed to finish.

Showing the debris impacted on its nose during an all-day race, the Moran/Benett C-4RK placed 10th at Le Mans in 1953. Its finishing average was 93.805 mph and its fastest lap was its sixth at 102.6 mph.

Although the coupe's chassis was identical to that of the roadster versions of the C-4R, many of its body features were unique. This included the louvered shrouding for its exhaust piping and the ducted exit just forward of the rear wheels. While its side windows were sliding at Le Mans, they were removed for racing in America.

Fitted with a three-spoke steering wheel made by Cunningham's West Palm Beach craftsmen, the C-4RK coupe had a similar dash layout with its electronic tachometer in the driver's line of sight. A fabric shroud kept its instrument lights from reflecting on the windshield and its special 24-hour clock for Le Mans was at the left.

By the time of the 1955 Grand Prix at Watkins Glen, New York the C-4RK had been acquired by Charles Moran. Racing it at the Glen was Chicagoan Fred Wacker, whose traditional "eight-ball" emblem it carried.

Fred Wacker finished eighth overall at Watkins Glen in 1955 in Moran's Delaware-registered C-4RK. Its truncated tail was evident, as were its roof-mounted fuel filler and its hatch for access to the spare tire.

Although Phil Walters drove the C-4RK on several more occasions in North America, including Elkhart Lake in 1952, he had come to "hate the car," according to aerodynamicist Ted Gondert, who said that Walters deliberately held it back at Elkhart to feign its inferiority. Cunningham would never again build a coupe for racing.

Watkins Glen in 1955 was the last racing appearance for the unloved C-4RK coupe. Although destined never to win a race, this magnificent Cunningham has justifiably gone down in history as one of the most aggressively exciting-looking sports-racing cars ever to take to the track.

The Cunningham team stunned Le Mans in 1953 with its new model, the C-5R. This had the seemingly retrograde feature of a solid tubular front axle guided by parallel radius rods. The entire front portion of its bodywork could easily be removed for attention to its internals. Its cylindrical oil cooler sat just in front of its water radiator.

For Phil Walters, leaning on the C-5R, 1953 was the year that Cunningham should have won Le Mans. However, he and his team fatally underestimated the speed capability of the C-Type Jaguars with their new disc brakes.

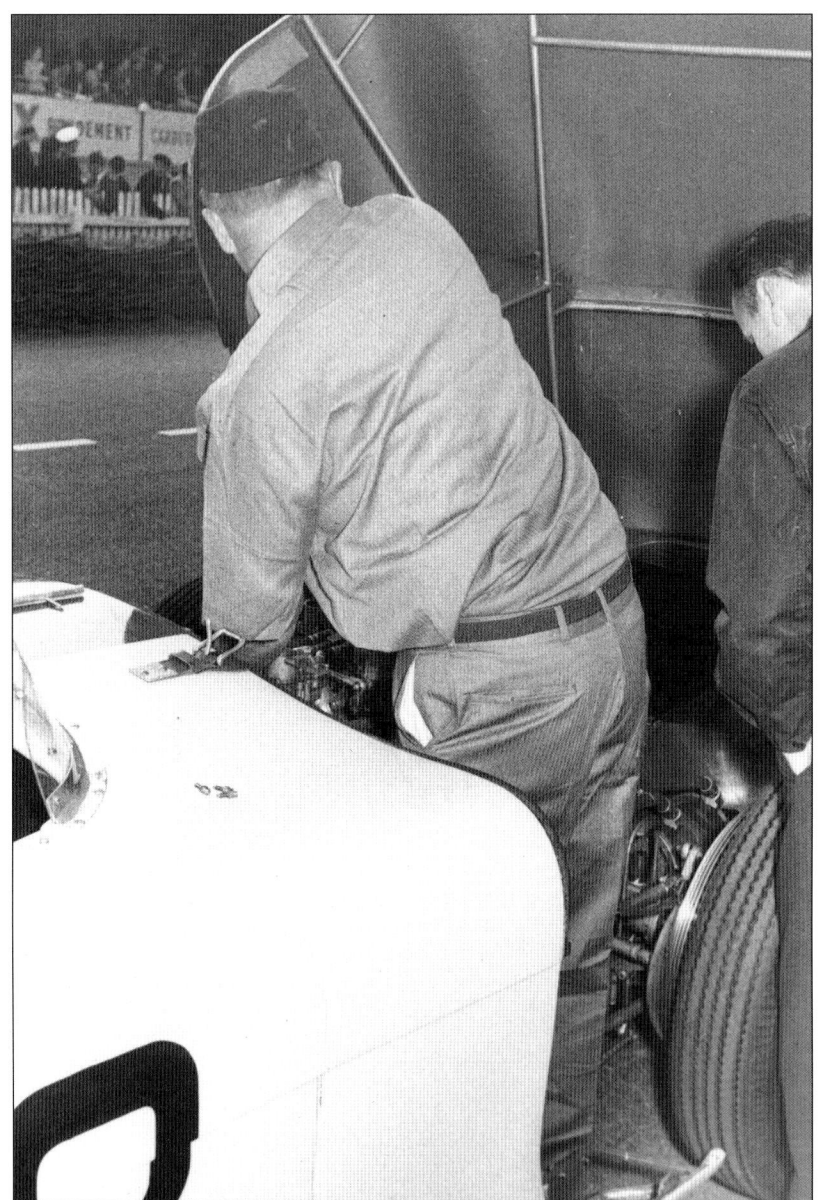 

A striking feature of the C-5R was the way its close-spaced tubular-steel frame rails allowed Cunningham's mechanics to stand right next to its engine for necessary attention. Circumferentially finned, its brake drums were a stunning 17 inches in diameter. No larger brakes have been fitted to any car in the modern era.

Removal of the drum exposed the cast-aluminum brake shoes of the C-5R. Following Ferrari practice of the time, each shoe was pivoted at its center and operated by cylinders at both ends. Automatic adjusters were provided as well. Hopes were high that these huge brakes would solve the problem of stopping consistently at Le Mans.

Crucial lateral guidance for the C-5R's solid front axle was provided by a vertical slide attached to the axle, seen at the left, which slid against a lubricated bronze block pivoted to the frame. Phil Walters characterized the system as a "de Dion axle at the front." Longitudinal torsion-bar springs were used at both ends of the C-5R.

Another Rudy Mailander photo of the C-5R's front end showed the long parallel radius rods locating the left side of the axle and the steel hoop and rubber buffer that limited suspension travel. Thanks to the axle's simplicity, said Phil Walters, it saved 30 pounds in unsprung weight over the independent front suspension of the C-4R Cunningham.

Sharing the track with two coupes, a Frazer-Nash and a Porsche, the C-5R was the fastest car at Le Mans in 1953. It was timed on the Mulsanne Straight at 154.81 mph and was capable of very quick lap times.

Although the new C-5R was first away at Le Mans in 1953 in the hands of Phil Walters, the cars ran to a strict schedule and were quickly overtaken by some of the more "racy" competitors who treated the first laps as a sprint race. An Allard and a supercharged Lancia coupe were coming up alongside the Cunninghams at the start.

Phil Walters and John Fitch (driving here) took three-hour stints in the C-5R in 1953. At the twelve-hour point in the race they were third behind a C-Type Jaguar and a Ferrari. With insufficiently compliant shoes, however, the big new drum brakes proved disappointing, shuddering and rumbling when used hard.

Sketches by Giovanni Michelotti, designer of Cunningham's C-3, provided the basis for the styling of the C-5R, whose body was the work of artist in aluminum Bob Blake. For obvious reasons the car was nicknamed "The Shark."

Fog lights blazing, John Fitch crossed the line to finish the 1953 24-hours in third place. However, because Charles Faroux, left, hadn't yet officially waved his flag, Fitch was chased up at the pits afterwards and reminded that he had to make one more lap to be classed as a finisher. His last lap was recorded as taking 27 minutes!

John Fitch and the C-5R crossed the line at Le Mans in 1953 just ahead of Duncan Hamilton in the winning C-Type Jaguar he shared with Tony Rolt. Inadequate venting of the Cunningham's under-hood air caused its hood side panels to be blown away. Louvers were later added to the side panels to overcome this problem.

Operating so far from its home base, the Cunningham team had to bring substantial resources with it to Le Mans. Derek Waller pictured the team preparing to depart from its base to the circuit in 1954 with two Chrysler sedans, a Cadillac, a truck and semi-trailer and a Bentley Continental in addition to the three team racing cars.

The three cars prepared by Cunningham for the Le Mans in 1954 included the two C-4Rs, one with a head-rest, and number 6, a Cunningham-bodied 4½-liter Ferrari. This was entered in lieu of the C-6R, which was not yet ready.

A glimpse inside the Cunningham storeroom at Le Mans in 1951 gave a hint of the magnitude of the supplies that the team brought to a still-straitened Europe. One observer likened the lavishness of the Cunningham equipment to the Berlin airlift. Operating far from home, the Cunningham team needed ample resources.

Also shipped to Europe for the Le Mans campaigns was Cunningham's mobile workshop, amply equipped with welding and machining equipment. Rudy Mailander photographed it at Le Mans in 1955, parked next to the 300SL-powered high-speed racing-car transporter used by the Mercedes-Benz team.

Apart from its hood badge, the Ferrari entered by the West Palm Beach team in 1954 was completely Cunningham-bodied. Surrounding its grille were fog lights, driving lights and air scoops for its liquid-cooled braking system.

Getting attention in the pits during night practice in 1954 at Le Mans, the Cunningham-entered Type 375 Ferrari used a system invented by Roy Sanford and developed by Raybestos-Manhattan that circulated cooling water through the brake shoes and two cooling radiators. Otherwise the chassis was standard Ferrari with a live rear axle.

The Cunningham people were understandably coy about the details of their brake-cooling system, which had two circulating pumps driven by the Ferrari's camshafts. Race preparer Alfred Momo, no stranger to Ferraris, fitted his own special external fuel reservoirs to the V-12 engine's three four-throat Weber carburetors.

Chief fabricator and mechanic Jack Donaldson, left, attended to some Ferrari details in 1954 while a capped Alfred Momo looked on. A supplementary instrument panel carried gauges for the brake-cooling system as well as a 24-hour clock. A vertical-column decelerometer to help judge braking intensity was directly in the driver's view.

In 1954 at Le Mans the Cunningham-bodied Ferrari was standing in for the team's new C-6R, which was delayed. Although its fastest lap of 4:26.0 was the quickest ever turned by a Cunningham entry at Le Mans, it took 10 seconds longer than the quickest lap of Ferrari's latest 4.9-liter V-12, the car that defeated a D-Type Jaguar to win.

Here passing a Frazer-Nash coupe, the Walters/Fitch Cunningham-Ferrari was sixth in the early running but then severely slowed by a rocker-arm breakage that reduced its effective cylinder count to 11. Never again higher than 20th, it retired at mid-race with rear-axle failure. It was timed at 153.43 mph on the Mulsanne Straight.

Briggs Cunningham's last sports-racing car, the C-6R, was finally ready for Le Mans in 1955. Bespectacled Alfred Momo stood behind the partially dismantled racer during pre-race preparations in the team's garage.

With John Fitch now driving for Mercedes-Benz and Phil Walters piloting his D-Type Jaguar, Briggs decided to handle his C-6R personally in 1955. With a Chrysler tender he drove it through the town's streets to the circuit.

Although designed by Briggs Weaver to be made as light-alloy castings, the front-suspension wishbones of the C-6R were fabricated of thin steel instead to eliminate any investment in patterns and molds. By 1955 it was evident that Briggs Cunningham was beginning to wind down his car-building operation and economies were inevitable.

The 13-inch drum brakes of the C-6R married a high-quality cast-iron braking surface to a hardened aluminum faceplate. Similar to the brakes on the C-4R as it was finally developed, these were very effective on a car weighing substantially less at 1,850 pounds dry. An approach to Dunlop for disc brakes had been snubbed.

Briggs Weaver designed a special input coupling to the C-6R's differential to absorb vibrations from its four-cylinder Meyer & Drake Offy engine. This replaced an earlier design, which used a small-diameter drive shaft that flexed in torsion. Inboard rear coil springs were operated by rocker arms pivoting on Timken roller bearings.

Fuel tankage on the 1955 C-6R married a tank above its differential with two side tanks along the car's sills. Its de Dion rear axle was controlled by parallel trailing radius rods and a vertical slot at the rear of the differential to provide lateral guidance. The central fuel tank's configuration made room for the spare wheel that the rules required.

The C-6R's rear drum brakes were adjacent to the differential as they had been on the original C-2 of 1951. The two-leading-shoe brakes had automatic adjustment and a sensor that showed how much lining life remained.

Fabrication of the C-6R's frame, fuel tank and oil tank gave an indication of the superb quality of workmanship that was available in West Palm Beach toward the end of the Cunningham car-building adventure. Just visible is the Jaguar six that was installed in the new chassis for racing in America after Le Mans in 1955.

As all-American as ever, the 1955 C-6R Cunningham was powered by an engine produced in Los Angeles, California by Meyer & Drake Engineering. Based on the 220-cubic-inch Offy engine used in sprint-car racing, Cunningham's four was reduced in stroke to bring its cylinder displacement under 183 cubic inches or three liters.

After experimentation with Hilborn fuel injection, twin Weber dual-throat carburetors were chosen as best for the Offy, burning gasoline instead of its usual alcohol. The Webers were flexibly mounted to reduce the effect of the engine's vibration on their mixture. Room for them was provided by the engine's 12-degree inclination to the left.

In Cunningham's Le Mans workshops the destroked Offy was overhauled in preparation for the 1955 race. Although the car raced in March at Sebring, little was learned before a half shaft broke. Le Mans would be the acid test for using this American track-racing engine, with its aluminum crankcase, in a road-racing car.

The Offy four had a cast-iron cylinder block with an integral head fed by four valves per cylinder. In the absence of the cooling effect of alcohol, problems were experienced with distortion of the inlet valves and seats that debilitated horsepower. Briggs Weaver developed special inlet valves with flexible tuliped heads to overcome this problem.

Ironically the problems experienced by the C-6R at Le Mans in 1955 had more to do with its ZF four-speed gear-box than with its Offy engine. Although the engine misfired initially, it later settled down to run strongly.

After 47 laps the C-6R made its first pit stop, lasting 28 minutes for attention to the gearbox, which early on had lost its lower ratios. Thus the car's timed maximum speed of 141.32 mph could not be considered representative.

Depleted of ratios though it was, the Cunningham C-6R soldiered on through the night and recovered to 13th place by the morning hours. Cunningham's D-Type Jaguar had long retired and the race was being run under the shadow of the catastrophe on the pit straight that cost 82 lives when a Mercedes-Benz 300SLR went into the crowd.

At Le Mans in 1955 Briggs Cunningham was partnered by tall Texan Sherwood Johnston, who preferred goggles to the visor favored by Briggs. If it was fated not to be the quickest Cunningham, the C-6R, with its stabilizing fin, stylish vents and scoops and elegant lines, certainly deserved its place in posterity as the prettiest Cunningham.

After completing three-quarters of the 1955 race, the C-6R finally retired in the 19th hour with a burned piston. This had been exacerbated by the struggle of Briggs and Johnston to lap the circuit with only top gear remaining. On Sunday morning on a rainy pit straight, the last of the Cunninghams was readied for its retirement.

Team manager Stanley Sedgwick had a word of consolation for Briggs Cunningham as the latter, in wet-weather garb, climbed the steps to his pit from his abandoned C-6R. Briggs and his team would achieve much more in racing through the early 1960s, but the epic car-building effort of the B. S. Cunningham Company was at an end.

More Great Titles From Iconografix

All Iconografix books are available from direct mail specialty book dealers and bookstores worldwide, or can be ordered from the publisher. For book trade and distribution information or to add your name to our mailing list and receive a **FREE CATALOG** contact:

Iconografix,
PO Box 446, Dept BK
Hudson, WI, 54016

Telephone: (715) 381-9755,
(800) 289-3504 (USA),
Fax: (715) 381-9756

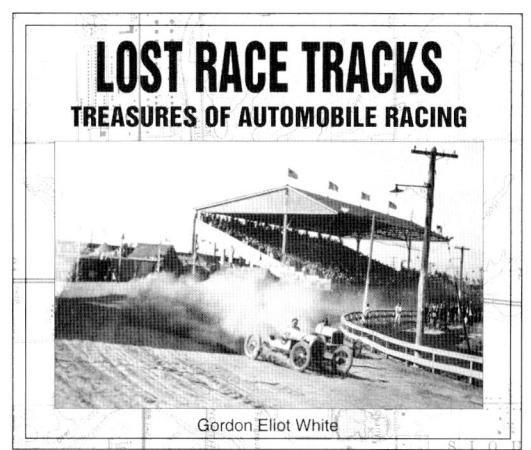

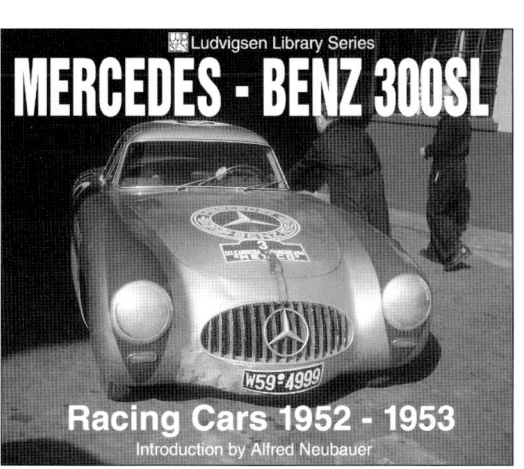

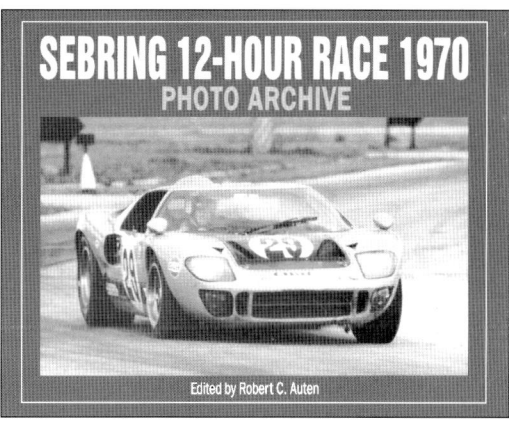

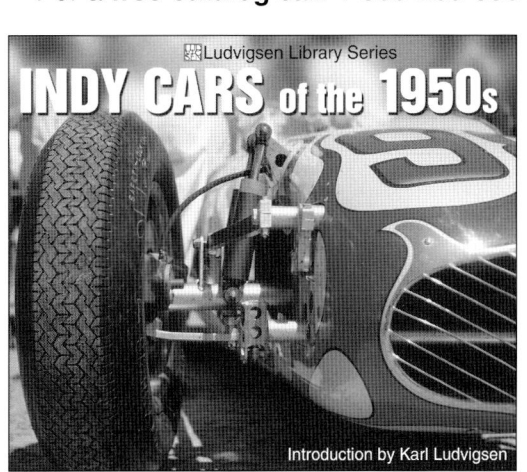

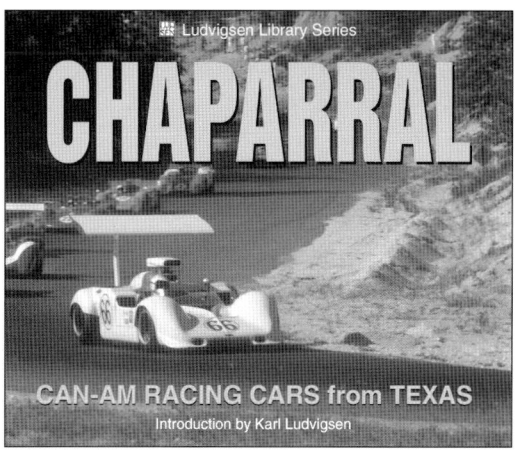

LUDVIGSEN LIBRARY